DOES THIS MAKE SENSE?
Constructing Explanations

Barbara A. Somervill

The Rosen Publishing Group's
PowerKids Press™
New York

For HPK, the greatest dad ever

Published in 2007 by The Rosen Publishing Group, Inc.
29 East 21st Street, New York, NY 10010

First Edition

Editor: Joanne Randolph
Book Design: Elana Davidian
Layout Design: Julio Gil

Photo Credits: Cover © Robert Manella/Getty Images; p. 4 © www.istockphoto.com/Alexei Naumov; p. 7 © Baumgartner Olivia/Corbis Sygma; p. 8 © James L. Stanfield/National Geographic Image Collection; p. 11 © Masterfile; p. 12 © PhotoDisc; p. 15 © British Antarctic Survey/Photo Researchers, Inc.; p. 16 © Dave Nagel/Getty Images; p. 19 © Kenneth Garrett/National Geographic Image Collection; p. 20 © www.istockphoto.com/Bonnie Jacobs.

Library of Congress Cataloging-in-Publication Data

Somervill, Barbara A.
    Does this make sense? : constructing explanations / Barbara A. Somervill.— 1st ed.
        p. cm. — (Think like a scientist)
    Includes bibliographical references and index.
    ISBN 1-4042-3485-3 (lib. bdg.) — ISBN 1-4042-2194-8 (pbk.)
    1. Science—Methodology—Juvenile literature. 2. Observation (Scientific method)—Evaluation—Juvenile literature.
3. Evidence—Evaluation—Juvenile literature. 4. Technical writing—Juvenile literature. I. Title. II. Series.
    Q175.2.S654 2007
    507.2—dc22
                                                2005035729

Manufactured in the United States of America

Contents

Crocodiles first existed about 200 million years ago. Scientists have studied fossils, or remains, of crocodiles from long ago and are able to see what has changed or stayed the same in crocodiles today.

The Scientific Method

Using the **scientific method** is an excellent way to prove an idea true. You begin with a question that needs answering. You make a guess that answers the question. That guess is called a **hypothesis**. This is the basis of your experiment. The scientific method is a step-by-step map that helps you find your way to a valid scientific conclusion.

People use the scientific method to explain how or why natural events occur. Why does fire sometimes help a forest? How have crocodiles changed since the days of dinosaurs? Why are there so many different types of sugar? As you follow the basic scientific method, you construct explanations. The **information** you collect provides **evidence** for these explanations.

What Is an Explanation?

Explanations tell you how to do something. They explain why something happens or give a reason for an event. You can use the information you collect in an experiment to answer questions and to make explanations.

Imagine you are baking cookies. The **recipe** calls for brown sugar, but you use powdered sugar instead. The cookies taste strange. Why? You decide to run an experiment to find out how different types of sugars have an effect on baked goods. You bake three batches of cookies using the same cookie recipe. You change the type of sugar in each batch, using brown sugar, white sugar, and powdered sugar. Your experiment will help you explain that different types of sugar change the taste and **texture** of foods.

Sugar comes in many forms. Some of the forms shown here are cubes of raw sugar, white sugar, powdered sugar, brown sugar, and rock candy, or crystallized sugar.

For an experiment on sugar, you will want to learn all you can about sugar. Did you know that sugar comes from sugar cane, shown here, as well as from a kind of vegetable called a sugar beet?

Using Data and Observations

As you conduct your experiment, you will collect plenty of data and **observations**. You will need to sort that information. You might sort the data by **category**, such as results and measurements. You do not need to include every bit of data you collect. Not every fact is important.

In your **research** you find out that brown sugar has molasses, but white and powdered sugar do not. The makeup of brown sugar is wetter and stickier than that of white or powdered sugar. You also find out that molasses is used in making taffy and other candy. What you learned in your research about brown sugar and molasses may help explain how different sugars have an effect on baked goods.

To help explain your experiment, you will want to use direct evidence. Direct evidence is easy to collect. It includes measurements, such as height, weight, mass, or volume. You will have collected direct evidence since your experiment began. You may **organize** direct evidence in tables, lists, graphs, and notes in your experiment log.

In the cookie experiment, you have an **ingredients** list for each batch of cookies. That is direct evidence. You might also make a **survey** that asks people their opinions about the taste and texture of each batch of cookies. The record of every step you took in doing the experiment and a copy of the survey are direct evidence of your work. You can use these records to explain how you ran your experiment.

Taking notes about the steps you take and what you learn as you do your experiment will help you when it comes time to explain your results.

Though it is not possible to use a ruler or tape measure to measure how far away another star is from Earth, scientists are able to use math and scientific theories to come up with this kind of data.

Indirect Evidence

Science often deals with **theories** that cannot be proved with direct evidence alone. Some theories are accepted based on a mix of direct and indirect evidence. These theories include how far away stars are and how those stars are formed.

The results of most experiments dealing with food are based on indirect evidence. You cannot measure flavor, or taste, with a ruler. You cannot weigh aroma. Some people really like sweets, and other people do not. People will also compare the cookies they taste in your test with cookies they have eaten in the past. They use that **experience** to form their opinions. The opinions you collect are based partly on indirect evidence, but they can be used in explaining your experiment.

Taking a Survey

In a science experiment, you are trying to explain your idea and the results of your experiment to others. The more facts you can gather, the better your explanation of a subject will be. Surveys are a way to collect data or test a hypothesis. Surveys use questions to collect information. They may collect information about certain facts, such as people's age or education. They may also collect opinions. Still other surveys do a count, like a **census**.

If you ask people to taste your cookies and give their opinions about the flavor, appearance, and texture of each cookie, you are conducting a survey. Your cookie survey has tasters rate each cookie on a scale of 1 to 10, with 10 being the top rating. You can use this information to explain how you decided which recipe produces the preferred cookie.

A scientist surveys the wandering albatross population of Bird Island. The information he collects on the number and size of eggs each bird lays will help him understand how we can keep this bird safe.

You might decide to get a random sample by choosing every other person who is waiting at your bus stop to be part of your survey.

Populations and Random Samples

Most surveys depend on populations and **random samples**. A population is a group. For example, it could be a group of people, animals, or trees. You can limit the population used in a survey. For example, you might ask only people with brown hair or only girls. You may present your population sample as 360 students or as 173 boys and 187 girls.

A random sample chooses a small group to stand for a large population. To get a random sample, pull names out of a hat or choose every third student in a line. The way you chose your taste testers needs to be included as part of the explanation of how you ran your experiment. It is direct evidence of the method you followed.

You need to explain your findings so that others can understand your experiment. You can use averages and ranges to help make your explanations.

You have asked three questions about each cookie. Add the ratings for each question then divide by the number of answers to get the average. If you have a rating of 10 and 8, the average would be 9. A range tells the upper and lower limits of the answers. The range on the cookie made with brown sugar was from 5 to 9.

Your survey results tell you that people liked the brown-sugar cookie the best overall. This means your explanation will say that the kinds of sugar used in baking change flavor, texture, and appearance.

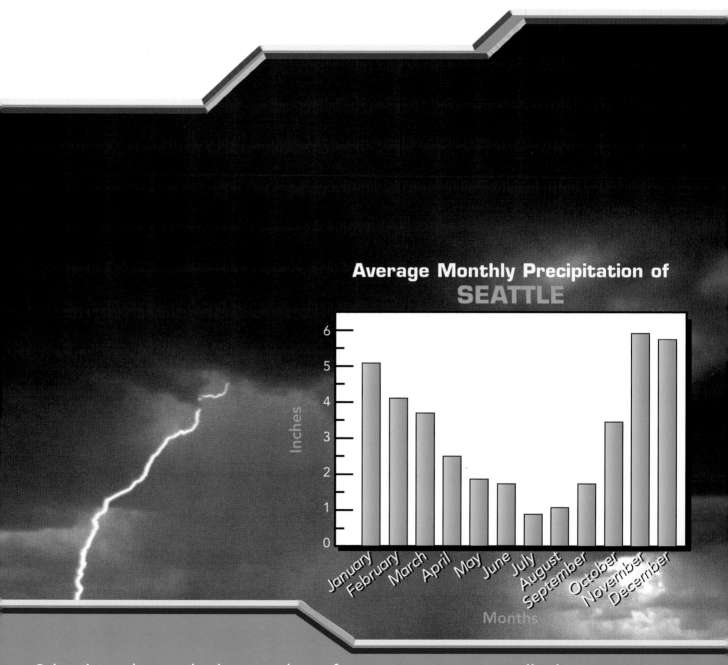

Average Monthly Precipitation of SEATTLE

Scientists who study the weather often use averages to talk about precipitation, or the amount of rain or snow that falls from the sky. This graph explains how much precipitation falls on average in Seattle.

When giving a presentation in front of the class, you may not want to go into as much detail as you do in your written report. You can summarize what happened and explain your results.

Creating a Summary

For your report you will want to include a short summary. Your summary gives a brief explanation of what you did and what you found out. Your report should tell the full story of your experiment, but the summary includes just the highlights. This summary could go at either the beginning or the end of your report. Scientists sometimes place their summaries, called **abstracts**, at the beginning of their reports.

The summary, or abstract, tells the reader what the report includes. In the cookie experiment, you might explain why and how you decided to do the experiment. An abstract serves as a brief explanation that helps people decide if they are interested in your experiment.

There are many ways to explain the information collected during an experiment. Each piece of information is evidence that will prove or disprove your hypothesis. Some information is direct evidence, such as measurements and the list of tools you used. Some information is indirect evidence.

Once you have all your evidence and information, you can use it to explain the results of your experiment. You will want to use clear and simple language in your explanation. State your hypothesis and then provide a step-by-step description of your experiment. Use pictures, charts, and graphic aids to help the reader understand your results. You want the reader to understand your experiment and your findings. Constructing a clear explanation is an important part of the scientific method.

Glossary

abstracts (AB-strakts) Brief summaries found at the front of scientific reports.

category (KA-teh-gor-ee) A group of things that are alike.

census (SEN-sus) The count of a population or group.

evidence (EH-vih-dints) Facts that prove something.

experience (ik-SPEER-ee-ents) Knowledge or skill gained by doing or seeing something.

hypothesis (hy-PAH-theh-ses) Something that is suggested to be true for the purpose of an experiment or argument.

information (in-fer-MAY-shun) Knowledge or facts.

ingredients (in-GREE-dee-unts) Parts of a combination.

observations (ahb-ser-VAY-shunz) Things that are seen or noticed.

organize (OR-guh-nyz) To have things neat and in order.

random (RAN-dum) Without a pattern.

recipe (REH-suh-pee) A set of directions for making something.

research (REE-serch) Careful study.

samples (SAM-pulz) Small parts of things, given to show what the rest is like.

scientific method (sy-un-TIH-fik MEH-thud) The system of running experiments in science.

survey (SER-vay) A set of questions asked of a number of people to find out what most people think about something.

texture (TEKS-chur) How something feels when you touch it.

theories (THEE-uh-reez) Ideas that try to explain something.

Index

Web Sites

Due to the changing nature of Internet links, PowerKids Press has developed an online list of Web sites related to the subject of this book. This site is updated regularly. Please use this link to access the list: www.powerkidslinks.com/usi/consexpl/